Pinpoint Math™

Student Booklet
Level E

Volume 1
Number Sense and Place Value

Photo Credits
©iStock International Inc., cover.

Acknowledgements
Content Consultant:

Linda Proudfit, Ph.D.

After earning a B.A. and M.A in Mathematics from the University of Northern Iowa, Linda Proudfit taught junior- and senior-high mathematics in Iowa. Following this, she earned a Ph.D. in Mathematics Education from Indiana University. She currently is Coordinator of Elementary Education and Professor of Mathematics Education at Governors State University in University Park, IL.

Dr. Proudfit has made numerous presentations at professional meetings at the local, state, and national levels. Her main research interests are problem solving and algebraic thinking.

www.WrightGroup.com

Copyright © 2009 by Wright Group/McGraw-Hill.

All rights reserved. Except as permitted under the United States Copyright Act, no part of this publication may be reproduced or distributed in any form or by any means, or stored in a database or retrieval system, without the prior written permission from the publisher, unless otherwise indicated.

Printed in USA.

Send all inquiries to:
Wright Group/McGraw-Hill
P.O. Box 812960
Chicago, IL 60681

ISBN 978-1-40-4568006
MHID 1-40-456800x

2 3 4 5 6 7 8 9 RHR 13 12 11 10 09

Contents

Tutorial Chart vii

Volume 1: Number Sense and Place Value
Topic 1 Place Value through 100
Topic 1 Introduction..1
Lesson 1-1 Tens and Ones 2–4
Lesson 1-2 Numbers to 100..5–7
Lesson 1-3 Compare and Order to 100 8–10
Lesson 1-4 Equivalent Forms 11–13
Topic 1 Summary...14
Topic 1 Mixed Review..15

Topic 2 Place Value through 1,000
Topic 2 Introduction..16
Lesson 2-1 Number to 1,000....................................17–19
Lesson 2-2 Write Numbers to 1,000............................20–22
Lesson 2-3 Compare and Order to 1,000 23–25
Topic 2 Summary...26
Topic 2 Mixed Review..27

Topic 3 Place Value beyond 1,000
Topic 3 Introduction..28
Lesson 3-1 Place Value to 10,000..............................29–31
Lesson 3-2 Expanded Notation with Zeros 32–34
Lesson 3-3 Numbers in the Millions............................35–37
Lesson 3-4 Round through Millions.............................38–40
Topic 3 Summary...41
Topic 3 Mixed Review..42

Glossary..43

Word Bank 44

Index 46

Objectives

Volume 1: Number Sense and Place Value

Topic 1 Place Value through 100

Lesson	Objective	Pages
Topic 1 Introduction	**1.2** Count, read, and write whole numbers to 100. **1.3** Compare and order whole numbers to 100 using the symbols for less than, equal to, or greater than (<, =, >). **1.4** Represent equivalent forms of the same number through the use of physical models, diagrams, and number expressions.	1
Lesson 1-1 Tens and Ones	**1.1** Count and group objects in ones and tens.	2–4
Lesson 1-2 Numbers to 100	**1.2** Count, read, and write whole numbers to 100.	5–7
Lesson 1-3 Compare and Order to 100	**1.3** Compare and order whole numbers to 100 using the symbols for less than, equal to, or greater than (<, =, >).	8–10
Lesson 1-4 Equivalent Forms	**1.4** Represent equivalent forms of the same number through the use of physical models, diagrams, and number expressions.	11–13
Topic 1 Summary	Review place value through 100.	14
Topic 1 Mixed Review	Maintain concepts and skills.	15

Topic 2 Place Value through 1,000

Lesson	Objective	Pages
Topic 2 Introduction	**2.1** Count, read, write whole numbers to 1,000 and identify the place value for each digit. **2.2** Use words, models, and expanded forms to represent numbers to 1,000.	16
Lesson 2-1 Numbers to 1,000	**2.1** Count, read, and write whole numbers to 1,000 and identify the place value for each digit.	17–19
Lesson 2-2 Write Numbers to 1,000	**2.2** Use words, models, and expanded forms to represent numbers to 1,000.	20–22
Lesson 2-3 Compare and Order to 1,000	**2.3** Order and compare whole numbers to 1,000 by using the symbols <, =, >.	23–25
Topic 2 Summary	Review place value through 1,000.	26
Topic 2 Mixed Review	Maintain concepts and skills.	27

Topic 3 Place Value beyond 1,000

Lesson	Objective	Pages
Topic 3 Introduction	**3.1** Identify the place value for each digit in numbers to 10,000. **3.2** Use expanded notation to represent numbers. **3.4** Round whole numbers through the millions to the nearest ten, hundred, thousand, ten thousand, or hundred thousand.	28
Lesson 3-1 Place Value to 10,000	**3.1** Identify the place value for each digit in numbers to 10,000.	29–31
Lesson 3-2 Expanded Notation with Zeros	**3.2** Use expanded notation to represent numbers.	32–34
Lesson 3-3 Numbers in the Millions	**3.3** Read and write whole numbers in the millions.	35–37
Lesson 3-4 Round through Millions	**3.4** Round whole numbers through the millions to the nearest ten, hundred, thousand, ten thousand, or hundred thousand.	38–40
Topic 3 Summary	Review skills related to place value beyond 1,000.	41
Topic 3 Mixed Review	Maintain concepts and skills.	42

Tutorial Guide

Each of the standards listed below has at least one animated tutorial for students to use with the lesson that matches the objective. If you are using the electronic components of *Pinpoint Math*, you will find a complete listing of Tutorial codes and titles when you access them either online or via CD-ROM.

Level E

Standards by topic	Tutorial codes
Volume 1 Number Sense and Place Value	
Topic 1 Place Value through 100	
1.1 Count and group objects in ones and tens.	1a Modeling Numbers with Base Ten Blocks
1.2 Count, read, and write whole numbers to 100.	1b Working with Whole Numbers, 0 to 100
1.3 Compare and order whole numbers to 100 using the symbols for less than, equal to, or greater than (<, =, >).	1c Ordering Whole Numbers, Example A
1.4 Represent equivalent forms of the same number through the use of physical models, diagrams, and number expressions.	1d Using Equivalent Forms to Represent Numbers, Example A
1.4 Represent equivalent forms of the same number through the use of physical models, diagrams, and number expressions.	1e Using Equivalent Forms to Represent Numbers, Example B
Topic 2 Place Value through 1,000	
2.1 Count, read, and write whole numbers to 1,000 and identify the place value for each digit.	2a Identifying the Place Value of Each Digit in a Whole Number, Example A
2.1 Count, read, and write whole numbers to 1,000 and identify the place value for each digit.	2b Ordering Whole Numbers, Example A
2.1 Count, read, and write whole numbers to 1,000 and identify the place value for each digit.	2c Working with Whole Numbers, 0 to 100
2.2 Use words, models, and expanded forms to represent numbers to 1,000.	2d Modeling Numbers with Base Ten Blocks
2.3 Order and compare whole numbers to 1,000 by using the symbols <, =, >.	2e Comparing Whole Numbers
2.3 Order and compare whole numbers to 1,000 by using the symbols <, =, >.	2b Ordering Whole Numbers, Example B
Topic 3 Place Value beyond 1,000	
3.1 Identify the place value for each digit in numbers to 10,000.	3a Identifying the Place Value of Each Digit in a Whole Number, Example B
3.1 Identify the place value for each digit in numbers to 10,000.	3b Writing Numbers in Expanded Form
3.2 Use expanded notation to represent four-digit numbers.	3b Writing Numbers in Expanded Form
3.3 Read and write whole numbers in the millions.	3c Writing Numbers in Word Form
3.4 Round whole numbers through the millions to the nearest ten, hundred, thousand, ten thousand, or hundred thousand.	3d Rounding

Topic 1: Place Value through 100

Topic Introduction

Complete with teacher help if needed.

1. Find the number.

   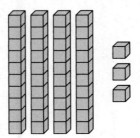

   a. _____ tens and _____ ones are shown.

   b. The number is _____.

   Objective 1.2: Count, read, and write whole numbers to 100.

2. Compare 34 and 37.

   a. 34 is _____ tens and _____ ones.

   b. 37 is _____ tens and _____ ones.

   c. _____ is the greater number because _____ _____

   Objective 1.3: Compare and order whole numbers to 100 by using the symbols for less than, equal to, or greater than.

3. Use MathFlaps™ to find each answer.

   a. 7 white + 9 blue = _____ in all

   b. 9 blue is _____ more than 7 white.

   Objective 1.4: Represent equivalent forms of the same number through the use of physical models, diagrams, and number expressions.

4. Compare 62 and 64.

   a. 62 is _____ tens and _____ ones.

   b. 64 is _____ tens and _____ ones.

   c. _____ is the greater number because _____ _____

   Objective 1.3: Compare and order whole numbers to 100 by using the symbols for less than, equal to, or greater than.

Lesson 1-1 — Tens and Ones

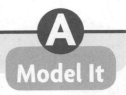

Words to Know We use these ten **digits** to make numbers.
0   1   2   3   4   5   6   7   8   9

Activity 1

Use base ten blocks. Show 43.

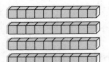

4 tens   3 ones

Practice 1

Use base ten blocks. Show 58. Write the number of tens and ones.

_____ tens _____ ones

Activity 2

How many blocks?

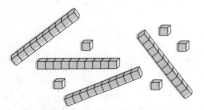

Count the tens. Count the ones. Write the number.

4 tens + 5 ones = 45

Practice 2

How many blocks? Count the tens. Count the ones. Write the number.

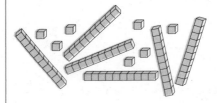

_____ tens + _____ ones = _____

On Your Own

Trade 10 ones for a ten. Then write the number.

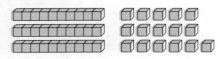

_____ tens + _____ ones = _____

Write About It

What do the numbers 38 and 83 have in common? Why are they not the same number?

Objective 1.1: Count and group objects in ones and tens.

Lesson 1-1 — Tens and Ones

B Understand It

Words to Know Each digit in a number has a **place**.
Each place has a **value**.
The value of the tens place is 10 times the digit.
The value of the ones place is 1 times the digit.

tens place → 7 in 72
ones place → 2 in 72

7 tens = 70   2 ones = 2

Example 1

Use the digits 5 and 7 to make four different two-digit numbers. Which number is the greatest? Which is the least?

tens	ones		tens	ones		tens	ones		tens	ones
5	7		7	5		5	5		7	7

Place-value charts show the values of the digits. If 5 is in the tens place, its value is 5 tens, or 50.

77 is the greatest number. 55 is the least.

Practice 1

Use the digits 2, 4, and 6 to make four different two-digit numbers.

tens	ones		tens	ones		tens	ones		tens	ones

Which number is the greatest? The least?

_____

Example 2

Write four different two-digit numbers that have 7 in the ones place.

47   17   87   37

Order the numbers from greatest to least.

87, 47, 37, 17

Practice 2

Write four different two-digit numbers that have 5 in the tens place.

_____

Order the numbers from least to greatest.

_____

On Your Own

Use the digits 3, 4, and 5. Write all the two-digit numbers that have a tens digit greater than the ones digit.

_____

Write About It

In what place is the digit 6 in the number 61? What is the value of the digit 6 in this number?

Objective 1.1: Count and group objects in ones and tens.

Lesson 1-1   **Tens and Ones**

1. How many blocks are there? Count the tens. Count the ones. Write the number.

   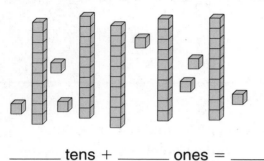

   _____ tens + _____ ones = _____

2. Write the tens and ones in each number.

   a. 64 = _____ tens + _____ ones

   b. 28 = _____ tens + _____ ones

   c. 90 = _____ tens + _____ ones

   d. 33 = _____ tens + _____ ones

3. List all the digits we use to make numbers.
   _____

4. Write four different two-digit numbers that have 6 in the ones place.
   _____

5. What is the tens digit in 47? Circle the letter of the correct answer.

   A 4   B 7   C 40   D 44

6. A two-digit number is less than 50. Which digit cannot be in the tens place? Circle the letter of the correct answer.

   A 2   B 3   C 4   D 7

7. Which of these means 7 tens and 4 ones? Circle the letter of the correct answer.

   A 47

   B 11

   C 74

   D 70

8. Write a two-digit number with a tens digit less than 4 and a ones digit greater than 6.
   _____

9. Angela switched the tens and ones digits in a number, but the number didn't change. If Angela's number has two digits, what could it be? Write all the answers.
   _____

10. What is the greatest two-digit number that you can make if you use only digits from 4 to 8?
    _____

Objective 1.1: Count and group objects in ones and tens.

Lesson 1-2: Numbers to 100

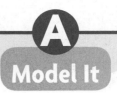

Words to Know

ones   tens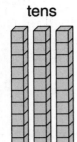

| 11 eleven | 12 twelve | 13 thirteen | 14 fourteen | 15 fifteen |
| 16 sixteen | 17 seventeen | 18 eighteen | 19 nineteen | 20 twenty |

Activity 1

Use base ten blocks to show 13.

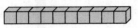

1 ten + 3 ones
10 + 3 = 13
           thirteen

Practice 1

Use base ten blocks to show 17.

Write the word form of 17.

_____

Activity 2

To count forward on a number line, move to the right. The ▼ shows 16.

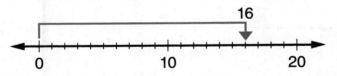

The word form of 16 is *sixteen*.

Practice 2

Show 18 on the number line.

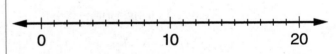

Write the word form of 18.

_____

On Your Own

Draw a number line from 0 to 20. Start at twelve and mark the next 4 numbers on the line.

Write About It

Gayle wrote the word name for 17 as *ten-seven*. Why is this a logical mistake?

Objective 1.2: Count, read, and write whole numbers to 100.

Lesson 1-2 Numbers to 100

B Understand It

Words to Know

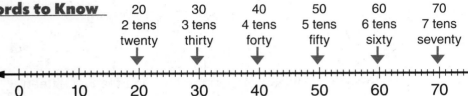

20	30	40	50	60	70	80	90	100
2 tens	3 tens	4 tens	5 tens	6 tens	7 tens	8 tens	9 tens	10 tens
twenty	thirty	forty	fifty	sixty	seventy	eighty	ninety	one hundred

Example 1

These tens and ones blocks show 45.

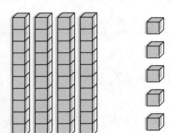

4 tens + 5 ones
40 + 5 = 45
forty-five

Practice 1

Write the number shown.
Then write the next 4 numbers.

_____ _____ _____ _____ _____

Example 2

Write the word name for the tens, then a hyphen, and then the word name for the ones.

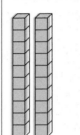

seventy-two 72

Practice 2

Write the word form of the number shown on the number line.

_____

On Your Own

Write the missing numbers.

67, 68, _____, _____, _____, 72

Write About It

When you count by tens, what changes? What stays the same?

Objective 1.2: Count, read, and write whole numbers to 100.

Volume 1   6   Level E

Lesson 1-2 — **Numbers to 100**

1. Show 36 in two different ways. Shade 36 blocks. Then mark the number line.

   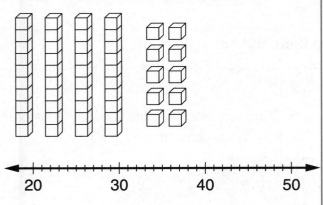

2. Write the missing numbers.

   a. 13, 14, _____, _____, _____, 18

   b. 47, 48, _____, _____, _____, 52

   c. 22, 23, _____, _____, _____, 27

   d. 75, 76, _____, _____, _____, 80

   e. 38, 39, _____, _____, _____, 43

3. Write the word names.

   a. 93 _____

   b. 15 _____

   c. 27 _____

   d. 49 _____

4. Which number is sixty-eight?

   A 66   B 68

   C 86   D 608

5. Which numbers between 0 and 100 have the same number of tens and ones?

   _____

6. Start at the number shown and count backward.

   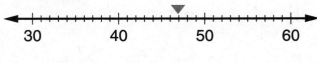

   ____ ____ ____ ____ ____

7. The word name for 14 uses the word *four*. Which other numbers from 11 to 19 include the ones word in their names?

   _____

8. On another sheet of paper, use a number line to count by 10s from 27 to 87.

   _____

   _____

Objective 1.2: Count, read, and write whole numbers to 100.

Lesson 1-3 — Compare and Order to 1,000

Words to Know Use the symbols < and > to **compare** two unequal numbers.

is less than   is greater than
    3 < 5   5 > 3

These statements are called **inequalities**.

Activity

Use base ten blocks to compare 347 and 265.

Show each number with hundreds, tens, and ones blocks.

347 = 3 hundreds 4 tens 7 ones

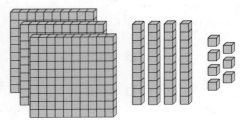

265 = 2 hundreds 6 tens 5 ones

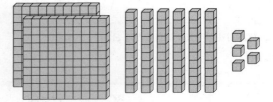

Start at the left and compare the hundreds.

3 hundreds > 2 hundreds, so 347 > 265.

Practice

Use base ten blocks to compare 446 and 464. The first model is drawn for you.

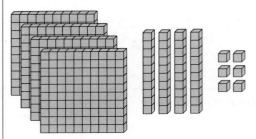

Draw a model for 464.

Start at the left. The hundreds are the same, so compare the tens. Use >, <, or =.

446 _____ 464

On Your Own

Complete. Use >, <, or =.

710 _____ 803   397 _____ 379

Write About It

How do you compare two 3-digit numbers when they have the same hundreds and tens digits? Give an example.

Objective 1.3: Order and compare whole numbers to 1,000 by using the symbols <, =, >.

Lesson 1-3: Compare and Order to 1,000

B Understand It

Words to Know   0   1   2   3   4   5   6   7   8   9

The number 9 is the **greatest**, or highest, number of this list.
The number 0 is the **least**, or lowest, number of this list.

Example

Order 604, 59, and 783 from least to greatest.

Use a place-value chart.

hundreds	tens	ones
6	0	4
	5	9
7	8	3

59 has 0 hundreds, so it is the least.
Compare 604 and 783.
Since $6 < 7$, 604 is less.

59, 604, 783

Practice

Order 516, 65, and 651 from greatest to least.

Start by writing the numbers in the place-value chart.

hundreds	tens	ones

_____ , _____ , _____

On Your Own

Mark these numbers on the number line. Put them in order from least to greatest.

602, 485, 845, 270, 671

_____ , _____ , _____ , _____ , _____

Write About It

Mario is ordering 367, 182, 65, and 207. He says 65 is the greatest. What mistake did he make?

Objective 1.3: Order and compare whole numbers to 1,000 by using the symbols $<$, $=$, $>$.

Lesson 1-3: Compare and Order to 1,000

Try It

1. Write in order from greatest to least.

   a. 194, 409, 419

   _____

   b. 256, 621, 97, 152

   _____

2. Write in order from least to greatest.

   a. 307, 73, 730

   _____

   b. 250, 820, 580, 500

   _____

3. Complete the exercises below. Use >, <, or = to make the sentence true.

   a. 138 _____ 183   b. 921 _____ 721

   c. 559 _____ 595   d. 425 _____ 452

   e. 628 _____ 86   f. 357 _____ 573

4. Which inequality is true? Circle the letter of the correct answer.

   A 184 > 95   B 527 > 720

   C 950 < 840   D 422 < 402

5. Mark has 147 baseball cards. Sarah has 163. Jasmine has 174. Who has the greatest number of baseball cards?

   _____

6. What is the greatest number that can be made using the digits 6, 2, and 8? Use each digit only once.

   _____

7. Explain how to order 451, 781, and 526 by making three comparisons.

   _____

   _____

   _____

   _____

8. A rose bush is 145 centimeters tall. A small palm tree is 415 centimeters. A climbing vine is 775 centimeters. Which plant is taller than 500 centimeters?

   _____

Objective 1.3: Order and compare whole numbers to 1,000 by using the symbols <, =, >.

Lesson 1-4 — Equivalent Forms

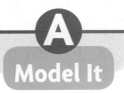

Words to Know

Pictures, models, symbols, words, and sums are **equivalent** if they represent the same number.

Names for Five

2 + 3   five   5   1 + 4

Activity 1

Show 12 as a sum in 4 different ways.

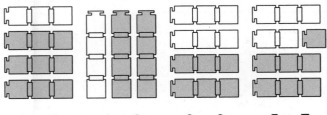

3 + 9   4 + 8   6 + 6   5 + 7

Practice 1

Use MathFlaps. Show 8 as a sum in 3 different ways. Draw pictures and write the sums.

_____   _____   _____

Activity 2

Write all the sums that equal 7.
Use 7 MathFlaps. Rearrange them in two different groups as many ways as possible.

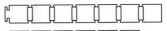   7 + 0
   6 + 1
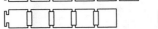   5 + 2
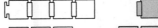   4 + 3
   3 + 4
   2 + 5
    1 + 6
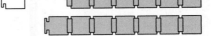   0 + 7

Practice 2

Write all the sums that equal 11. Use MathFlaps if you need help finding them all.

_____ _____ _____ _____

_____ _____ _____ _____

_____ _____ _____ _____

On Your Own

Write all the sums that equal 6.

_____

_____

Write About It

Alexis wrote 15 as 5 + 10, 6 + 9, and 7 + 8. How can she use these sums to get three more names for 15?

Objective 1.4: Represent equivalent forms of the same number through the use of physical models, diagrams, and number expressions.

| Lesson 1-4 | Equivalent Forms |  Understand It |

Words to Know Use **tally marks** to show groups of 5. Mark 4 lines. Cross them with a fifth line.

Example 1

Show 9 as a sum two different ways. Use a number line.

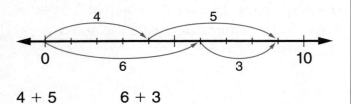

4 + 5   6 + 3

Show 13 as a sum two different ways. Use a number line.

Example 2

Use tally marks to show 14.

Practice 2

Use tally marks to show 18.

On Your Own

Show 10 as a sum by drawing two different models. You can draw pictures, number lines, or tally marks.

Write About It

Show the number 17 with MathFlaps and on a number line. Draw both models.

Is it easier to use objects or a number line for finding all the different ways to make a sum of 17?

Objective 1.4: Represent equivalent forms of the same number through the use of physical models, diagrams, and number expressions.

Lesson 1-4   **Equivalent Forms**

Try It

1. Shade to show 4 different ways to make a sum of 9.

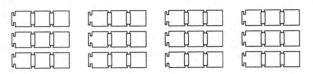

2. Show 2 different ways to make a sum of 12.

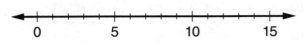

3. Write 3 different sums that equal each number.

   a. 14 _____

   b. 13 _____

   c. 10 _____

4. What is another name for 7?

   A 0 + 7   **B** 5 + 6

   C 7 + 2   **D** 10 + 7

5. Look at the tally marks. Using numerals, write 4 different names for the sum.

   _____

   _____

6. Use subtraction. Show four different names for 3.

   _____

   _____

7. Margie is adding 8 to 17. Why is it helpful to write 17 as 2 + 15?

   _____

   _____

8. An egg carton holds 1 dozen eggs. What sums can you model using an entire egg carton? Draw one model with your answer.

Objective 1.4: Represent equivalent forms of the same number through the use of physical models, diagrams, and number expressions.

Topic 1: Place Value through 100

Topic Summary

Circle the letter of the correct answer. Explain how you decided.

1. What number does the model show?

   A 22

   B 25

   C 27

   D 72

   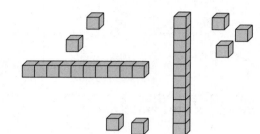

   _____
   _____
   _____

2. Which is greater, 14 − 8 or 4 × 2? Use the models to help you decide.

   A 14 − 8

   B 4 × 2

   C They are equal.

   _____
   _____
   _____

Objective: Review place value through 100.

Topic 1: Place Value through 100

Mixed Review

1. Compare 88 and 89.

  a. 88 is _____ tens and _____ ones.

  b. 89 is _____ tens and _____ ones.

  c. _____ is greater because _____

  _____

Volume 1, Lesson 1-3

2. Write the word form for each number.

  a. 27 _____

  b. 8 _____

  c. 66 _____

  d. 50 _____

Volume 1, Lesson 1-2

3. Which comparison is NOT correct? Circle the letter of the correct answer.

  A 97 > 7

  B 98 > 89

  C 2 < 21

  D 25 < 24

Volume 1, Lesson 1-3

4. Write four different numbers using 1, 4, 7, and 9 that have 4 in the tens place.

  _____

Volume 1, Lesson 1-2

Objective: Maintain concepts and skills.

Topic 2: Place Value through 1,000

Topic Introduction

Complete with teacher help if needed.

1.

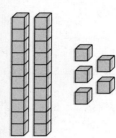

a. _____ tens are shown.

b. _____ ones are shown.

c. The number in numerals is _____.

d. The number in words is

_____.

Objective 2.2: Use words, models, and expanded forms to represent numbers to 1,000.

---

2. 3 7 1

a. The number in words is

_____.

b. There are _____ ones.

c. There are _____ tens.

d. There are _____ hundreds.

Objective 2.1: Count, read, and write whole numbers to 1,000 and identify the place value for each digit.

---

3. 8 9 4

a. The number in words is

_____.

b. _____ is in the tens place.

c. _____ is in the hundreds place.

d. _____ is in the ones place.

Objective 2.1: Count, read, and write whole numbers to 1,000 and identify the place value for each digit.

---

4. 2 6 7

a. The number in words is

_____.

b. 7 is in the _____ place.

c. 2 is in the _____ place.

d. 6 is in the _____ place.

Objective 2.1: Count, read, and write whole numbers to 1,000 and identify the place value for each digit.

Lesson 2-1 | **Numbers to 1,000**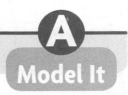

Activity 1

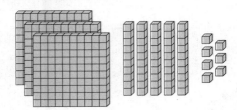

These base ten blocks show
3 hundreds, 5 tens, and 7 ones.

Write the number in the chart.

hundreds	tens	ones
3	5	7

= 357

Practice 1

Use base ten blocks to show 293.
Draw a picture of your model.

Write your number in the chart.

hundreds	tens	ones

Activity 2

Look at the blocks above. Count by hundreds.

   100,   200,   500

Now count on by tens.

   310,   320,   330,   340,   350

Now count on by ones.

   351,   352,   353,   354,   355,   356,   357

Practice 2

Look at your model in practice 1 above.

Count the hundreds. _____, _____,

Count by tens. _____, _____, _____,

_____, _____, _____,

_____, _____, _____

Count on by ones. _____, _____, _____

On Your Own

Write the number that has 1 more 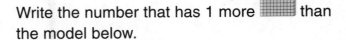 than the model below.

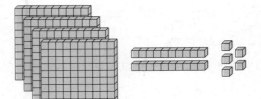

Write About It

Show how to move the base ten blocks to make the number easier to name. Explain.

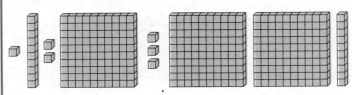

Objective 2.1: Count, read, and write whole numbers to 1,000 and identify the place value for each digit.

Lesson 2-1: Numbers to 1,000

 Understand It

Words to Know   hundreds place   tens place   ones place

7   4   9

Example 1

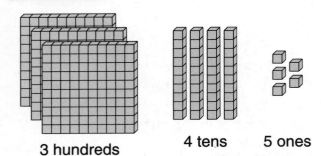

3 hundreds   4 tens   5 ones

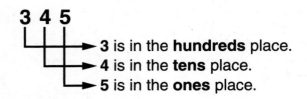

- 3 is in the **hundreds** place.
- 4 is in the **tens** place.
- 5 is in the **ones** place.

Practice 1

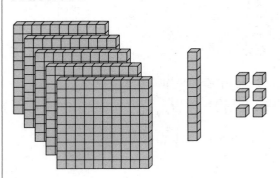

_____ is in the hundreds place.

_____ is in the tens place.

_____ is in the ones place.

Example 2

Expanded form: 400 + 70 + 5

Word form: four hundred seventy-five

Standard form: 475

Model:

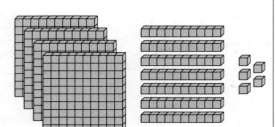

Practice 2

Expanded form: 500 + 60 + 3

Word form: _____

Standard form: _____

Model:

On Your Own

What number has 5 ones, 3 hundreds, and 2 tens?

_____

Write About It

Compare the expanded forms of 641 and 849.

Objective 2.1: Count, read, and write whole numbers to 1,000 and identify the place value for each digit.

Lesson 2-1 | **Numbers to 1,000**

1. Write the number the model shows.

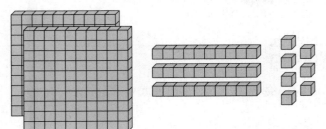

   a. _____

   Write the number for each place.

   b. _____ is in the hundreds place.

   c. _____ is in the tens place.

   d. _____ is in the ones place.

2. Give the expanded and word forms for each number.

   a. 874 _____
   _____

   b. 193 _____
   _____

   c. 562 _____
   _____

3. Name the place.

   a. In 786, 6 is in the _____ place.

   b. In 457, 4 is in the _____ place.

   c. In 193, 9 is in the _____ place.

4. Circle the letter of the number whose word form is *six hundred fifty-eight*.

   A 685   B 658

   C 586   D 568

5. Write two different numbers. Use 2, 5, and 7 and have 2 in the tens place.

   _____   _____

6. Fill in the blanks. Count by tens.

   ____, ____, ____, 360, ____,

   ____, ____

7. In a three-digit number written in expanded form, how many different numbers can be in the middle? What are they?

   _____
   _____

8. A number has 7 in the ones place, 9 in the hundreds place, and 6 in the tens place. What is the word form of the number?

   _____

Objective 2.1: Count, read, and write whole numbers to 1,000 and identify the place value for each digit.

Lesson 2-2: Write Numbers to 1,000

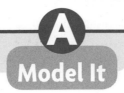
Model It

Words to Know
The **standard form** of a number: 359
The **expanded form** of a number: 300 + 50 + 9
The **word form** of a number: *three hundred fifty-nine*

Activity

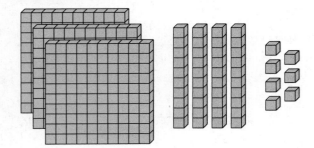

Here are three ways to write the number modeled above.

Standard form: 347
Expanded form: 300 + 40 + 7
Word form: *three hundred forty-seven*

Practice

Use base ten blocks to show 254.

Write the number in expanded form.
_____

Write the number in word form.
_____

On Your Own

Count up by tens from 542 using expanded form. Then show the standard form of each number.

500 + 40 + 2 = 542

____ + ____ + ____ = ____

____ + ____ + ____ = ____

____ + ____ + ____ = ____

Write About It

The expanded form of 618 is 600 + 10 + 8. Tell why you think the word *expanded* describes this form.

Objective 2.2: Use words, models, and expanded forms to represent numbers to 1,000.

Lesson 2-2 — Write Numbers to 1,000

Understand It — B

Example 1

Expanded form: 400 + 70 + 5

Word form: four hundred seventy-five

Standard form: 475

Model:

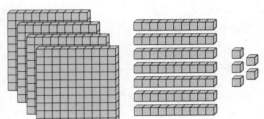

Practice 1

Expanded form: 500 + 60 + 3

Word form: _____

Standard form: _____

Model:

Example 2

These all name the same number.

346

3 hundreds, 4 tens, 6 ones

300 + 40 + 6

three hundred forty-six

Practice 2

Write the number 729 three ways.

_____ hundreds, _____ tens, _____ ones

Expanded form: _____ + _____ + _____

Word form: _____

On Your Own

Circle parts of the model to match the expanded form.

300 + 50 + 4

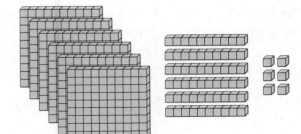

Write About It

Use the numbers 372 and 732.

How are the expanded forms of the numbers the same?

_____

_____

How are the expanded forms of the numbers different?

_____

_____

Objective 2.2: Use words, models, and expanded forms to represent numbers to 1,000.

Lesson 2-2 — Write Numbers to 1,000

Try It

1. Give the expanded and word forms for each number.

   a. 874 _____

   _____

   b. 193 _____

   _____

   c. 562 _____

   _____

2. Write the number modeled below in three different ways.

   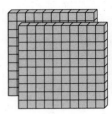 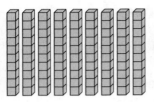

   Standard form _____

   Expanded form _____

   Word form _____

3. Write in standard form a number that has the given addend when written in expanded form.

   a. a tens place value of 40 _____

   b. a hundreds place value of 300 _____

   c. a ones place value of 3 _____

   d. a tens place value of 0 _____

4. What is the standard form of *six hundred fifty-eight*?

   A 685

   B 658

   C 586

   D 568

5. In a three-digit number written in expanded form, which numbers can be in the middle?

   _____

   _____

6. A number has 4 in the ones place, 9 in the hundreds place, and 5 in the tens place. What is the word form of the number?

   _____

   _____

Objective 2.2: Use words, models, and expanded forms to represent numbers to 1,000.

Lesson 2-3 — Compare and Order to 1,000

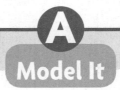
Model It — A

Words to Know Use the symbols <, =, > to **compare** two numbers.

is less than	is equal to	is greater than
3 < 5	8 = 8	10 > 6

Activity 1

Compare 236 and 118. Write >, <, or =.

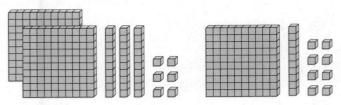

Compare the hundreds.
2 hundreds > 1 hundred, so 236 > 118

Practice 1

Use base ten blocks to compare 326 and 452.

Compare the hundreds.

_____ hundreds is _____ than _____ hundreds.

Write >, <, or =.

326 ◯ 452

Activity 2

Compare 520 and 580.
Mark the numbers on a number line.

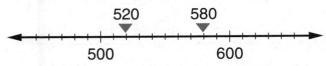

Compare the hundreds. They are the same.
Compare the tens. 2 tens < 8 tens
520 < 580

Practice 2

Compare 610 and 670.
Mark the numbers on a number line.

Write >, <, or =.

610 ◯ 670

On Your Own

Show two ways to compare 237 and 327.
Use >, <, or =.

237 ◯ 327   327 ◯ 237

Write About It

Two 3-digit numbers have the same number of hundreds. When are these numbers equal?

Objective 2.3: Compare and order whole numbers to 1,000 by using the symbols <, =, >.

Lesson 2-3 Compare and Order to 1,000

Words to Know   **Writing in order** or **ordering** means listing numbers from greatest to least or from least to greatest.

Example 1

Write 288, 382, and 223 in order from greatest to least.
Look at the hundreds.
Since 382 has 3 hundreds, it is greatest.
Then, 288 > 223 because 8 > 2.

382, 288, 223

Practice 1

Write the numbers below in order from greatest to least. Cross out each number as you use it.

413, 147, 417, 236

_____, _____, _____, _____

Example 2

Write 761, 630, 872, and 711 in order from least to greatest by placing them on a number line.

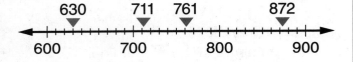

Practice 2

Write 479, 381, 522, and 321 in order from least to greatest by placing them on the number line.

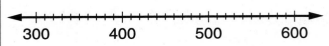

On Your Own

Use the number line. Write these numbers in order. 552, 483, 229, 840, 320, 711.

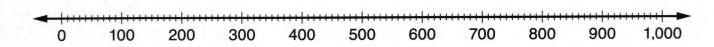

Write About It

What happens if you write 318, 542, 234, and 796 in order using just the ones? Why doesn't this work?

Objective 2.3: Compare and order whole numbers to 1,000 by using the symbols <, =, >.

Lesson 2-3 — **Compare and Order to 1,000**

Try It

1. Write in order from least to greatest.

   a. 466, 243, 408, 72

   b. 85, 519, 130, 935, 246

   c. 320, 852, 612, 274, 336

2. Write in order from greatest to least.

   a. 776, 773, 821, 654

   b. 652, 502, 765, 48, 517

   c. 601, 234, 452, 164, 578

3. Complete. Use >, <, or =.

   a. 409 _____ 109   b. 733 _____ 443

   c. 619 _____ 916   d. 520 _____ 679

   e. 305 _____ 852   f. 934 _____ 892

4. Which statement is true? Circle the letter of the correct answer.

   A 340 = 430   B 327 < 723

   C 518 > 658   D 901 < 109

5. Use 115, 324, 257, >, and <. Write as many different statements as you can.

6. Name three numbers that are less than 823 and greater than 805.

7. Paolo owned two paintings, one worth $830 and one worth $790. How will you find the painting that is worth more?

8. 939 is greater than 845, and 845 is greater than 608. Write two statements comparing 939 and 608. Use words.

Objective 2.3: Compare and order whole numbers to 1,000 by using the symbols <, =, >.

Topic 2: Place Value through 1,000

Topic Summary

Choose the correct answer. Explain how you decided.

1. Which of these has 4 tens, 3 ones, and 6 hundreds?

   A 436

   B 643

   C 364

   D 634

   _____
   _____
   _____

2. Which of these is the expanded form for five hundred thirteen?

   A 500 + 13

   B 5 + 1 + 3

   C 500 + 10 + 3

   D 500 + 3

   _____
   _____
   _____

Objective: Review place value through 1,000.

Topic 2: Place Value through 1,000

Mixed Review

1. Give the expanded form for each number.

   a. 632 _____

   b. 41 _____

   c. 857 _____

   Volume 1, Lesson 2-2

2. Compare. Use <, >, or =.

   a. 54 _____ 58

   b. 62 _____ 26

   c. 45 _____ 45

   Volume 1, Lesson 1-3

3. Write the word names for each number.

   a. 53 _____

   b. 609 _____

   c. 20 _____

   d. 328 _____

   Volume 1, Lesson 1-2

4. Show 10 as a sum in three different ways.

   _____

   Volume 1, Lesson 1-4

5. Tell the digit in each place.
   765

   a. tens _____

   b. ones _____

   c. hundreds _____

   Volume 1, Lesson 2-1

6. Write the word form and standard form for $200 + 5 + 9$.

   Word form: _____

   Standard form: _____

   Volume 1, Lesson 2-2

Objective: Maintain concepts and skills.

Topic 3: Place Value beyond 1,000

Topic Introduction

Complete with teacher help if needed.

1. 7,621

   a. The value of 2 is _____.

   b. The value of 6 is _____.

   c. The value of 7 is _____.

   d. The value of 1 is _____.

2. Write 19,702 in expanded notation.

   _____

Objective 3.1: Identify the place value for each digit in numbers to 10,000.

Objective 3.2: Use expanded notation to represent numbers.

3. 42,312

   a. _____ is in the tens place.

   b. _____ is in the hundreds place.

   c. _____ is in the ones place.

   d. _____ is in the thousands place.

   e. _____ is in the ten thousands place.

   f. Round to the nearest thousand. _____

4. 9,827,654

   a. 6 is in the _____ place.

   b. 9 is in the _____ place.

   c. 7 is in the _____ place.

   d. 8 is in the _____ place.

   e. 2 is in the _____ place.

   f. Round to the nearest ten thousand.

   _____

Objective 3.4: Round whole numbers through the millions to the nearest ten, hundred, thousand, ten thousand, or hundred thousands.

Objective 3.4: Round whole numbers through the millions to the nearest ten, hundred, thousand, ten thousand, or hundred thousand.

Volume 1 — Level E

Lesson 3-1: Place Value to 10,000

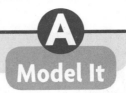

Model It

Activity 1

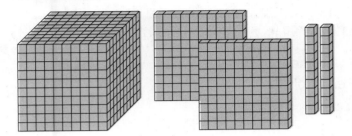

These base ten blocks show 1 thousand, 2 hundreds, 2 tens, and 0 ones.

thousands	hundreds	tens	ones
1	2	2	0

The number is 1,220.

Practice 1

Draw a model of the number 2,563. Then write the number in the chart.

thousands	hundreds	tens	ones

Activity 2

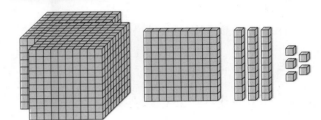

Standard form: 2,135
Expanded form: 2,000 + 100 + 30 + 5
Word form: two thousand, one hundred thirty-five

Practice 2

Draw a model of 4 thousands, 7 hundreds, 4 tens, and 3 ones.

Standard form _____

Expanded form _____

Word form _____

On Your Own

Use the digits 5, 4, 7, and 3. Make four-digit numbers where the greatest digit is in the greatest place value and the least digit is in the least place value. How many different numbers can you make? What are they?

Write About It

How would you represent the value of each 4 in the number 4,426?

Objective 3.1: Identify the place value for each digit in numbers to 10,000.

Lesson 3-1 Place Value to 10,000

Understand It

Words to Know The **value** of a digit is determined by the place it is in.
In the number 4,528, the **value** of the **2** is **2 tens**, or **20**.

Example 1

6,4 8 3

6 is in the **thousands place**.
4 is in the **hundreds place**.
8 is in the **tens place**.
3 is in the **ones place**.

Practice 1

Use the number 9,572. Tell which digit is in each place.

_____ is in the tens place.

_____ is in the thousands place.

_____ is in the ones place.

_____ is in the hundreds place.

Example 2

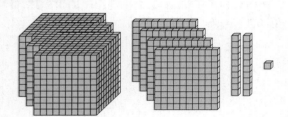

3 **thousands** have a value of **3,000**.
4 **hundreds** have a value of **400**.
2 **tens** have a value of **20**.
1 **one** has a value of **1**.

Practice 2

Use the number 6,513.

The value of the **3** is _____.

The value of the **5** is _____.

The value of the **6** is _____.

The value of the **1** is _____.

On Your Own

In the number 9,162, the 9 is in the thousands place. Its **value** is 9,000.

What is **the value of the 1?** _____

What is **the value of the 6?** _____

What is **the value of the 2?** _____

Write About It

How do the values of the digit 5 differ in the numbers 5,476 and 2,953? Explain.

Objective 3.1: Identify the place value for each digit in numbers to 10,000.

Lesson 3-1 — Place Value to 10,000

Try It

1. Write the standard form of the number modeled below. Then name the digit in each place.

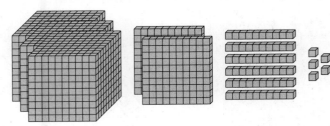

Standard form _____

_____ is in the hundreds place.

_____ is in the ones place.

_____ is in the thousands place.

_____ is in the tens place.

2. Give the value of the bold digit in each number.

a. 5,**4**38 _____

b. 9,1**3**2 _____

c. 7,6**8**5 _____

d. **4**,279 _____

e. 6,**4**91 _____

f. 3,745 _____

g. 9,8**2**7 _____

3. Give the value of the digit 9 in each number.

a. 8,952 _____   b. 9,163 _____

c. 7,495 _____   d. 2,419 _____

4. Circle the letters of all the numbers where the digit 4 has the value 4,000.

A 4,592   B 3,941

C 2,784   D 4,627

5. Write these numbers in expanded form.

a. 5,632 _____

b. 9,456 _____

6. Fill in the place-value chart for 7,591.

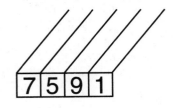

7. Write the word form of 6,195.

_____

_____

8. Give the place of the bold digit.

a. 9,**4**28 _____

b. **2**,317 _____

Objective 3.1: Identify the place value for each digit in numbers to 10,000.

Lesson 3-2 — Expanded Notation with Zeros

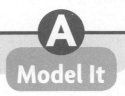

Activity 1

There are no tens in 3,402. Zero is in the tens place.

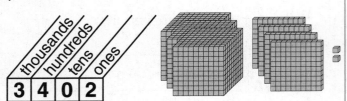

3 thousands, 4 hundreds, 0 tens, 2 ones

Practice 1

Use the chart to write the number shown.

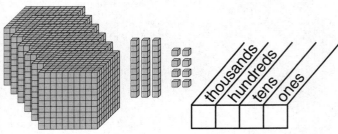

Which place has a 0 in it? _____

What does the 0 mean?
_____

Write the number of thousands, hundreds, tens and ones.
_____
_____

Activity 2

When a 0 digit occurs in a number, do not include that place value in the expanded notation.

4,907 = 4,000 + 900 + 7

Practice 2

Write 7,001 in a place-value chart. Then write its expanded notation.
_____

On Your Own

Write the following numbers in a place-value chart.   604   3,020   8,003

Which number(s) have 0 ones? _____

Which number(s) have 0 tens? _____

Write About It

How are the expanded notations for 6,927 and 6,027 different?

Objective 3.2: Use expanded notation to represent four-digit numbers.

Volume 1 — 32 — Level E

Lesson 3-2: Expanded Notation with Zeros

B Understand It

Example 1

Write the expanded form of the number shown in the place-value chart below. Then write the word form of the number.

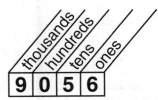

The expanded form of the number is 9,000 + 50 + 6. There are no hundreds.

The word form is *nine thousand, fifty-six*.

Practice 1

Write the expanded form of the number shown in the place-value chart below.

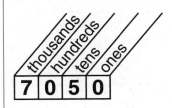

_____

Write the word form of the number.

_____

Example 2

Use the expanded form below.

Write the standard form of the number.

Then write the word form of the number.

5,000 + 30 + 7

The expanded form shown above means that the standard form of the number has a 0 in the hundreds place: 5,037.

The word form of the number is *five thousand, thirty-seven*.

Practice 2

Use the expanded form below.

8,000 + 300 + 9

Write the standard form of the number.

_____

Write the word form of the number.

_____

On Your Own

Write the expanded form of 4,368 after the number has been decreased by 6 tens.

_____

Write the word form of the new number.

_____

Write About It

What is the greatest number of zeros you could have in a three-digit number? Explain.

Objective 3.2: Use expanded notation to represent four-digit numbers.

Lesson 3-2 — **Expanded Notation with Zeros**

Try It

1. Write the expanded form of each number.

   a. 5,209 _____

   b. 730 _____

   c. 8,007 _____

   d. 1,036 _____

2. Write the standard form of each number.

   a. 300 + 50 _____

   b. 7,000 + 800 + 40 + 3 _____

   c. 2,000 + 5 _____

   d. 4,000 + 70 _____

3. Write the expanded form of the number shown in the place-value chart below.

   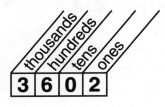

   _____

4. Janine drove 3,072 miles on Saturday. Circle the letter of the expanded form of 3,072.

   A 3,000 + 700 + 2

   B 3,000 + 200 + 7

   C 300 + 70 + 2

   D 3,000 + 70 + 2

5. Reverse the digits of the number 4,020. Is the new number still a four-digit number? Explain.

   _____
   _____
   _____

6. Reverse the digits of the number 7,805. Is the new number still a four-digit number? Explain.

   _____
   _____
   _____

7. The expanded form of a four-digit number has only two addends. What is the greatest number it could be? What is the least number?

   _____

8. Write all the numbers that include only 8,000 and a ten-place value in their expanded forms.

   _____
   _____
   _____

Objective 3.2: Use expanded notation to represent four-digit numbers.

Lesson 3-3: Numbers in the Millions

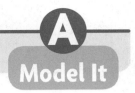

Words to Know

A **period** is a set of three place values.

The **ones period** includes the ones, tens and hundreds places.
The **thousands period** includes the thousands, ten thousands, and hundred thousands places.
The **millions period** includes the millions, ten millions, and hundred millions places.

Activity 1

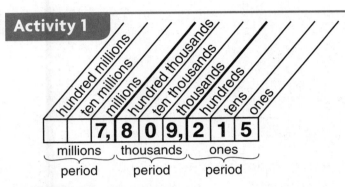

Standard form: 7,809,215

Word form: *seven million, eight hundred nine thousand, two hundred fifteen*

Each period except the ones period ends with the name of that period followed by a comma.

Practice 1

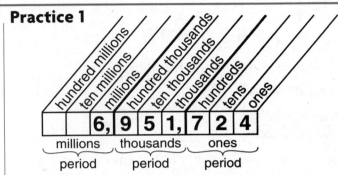

Write the word form of the number.

_____

_____

_____

_____

Activity 2

Write **5,950,306** in word form.

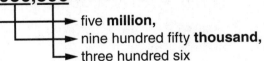

→ five **million,**
→ nine hundred fifty **thousand,**
→ three hundred six

Practice 2

Write 8,042,507 in word form.

_____

_____

_____

On Your Own

Write a seven-digit number that does not repeat any digits. Then write its word form.

_____

_____

_____

Write About It

How can you tell where the commas are placed in a large number?

Objective 3.3: Read and write whole numbers in the millions.

Lesson 3-3 — Numbers in the Millions

Understand It

Example 1

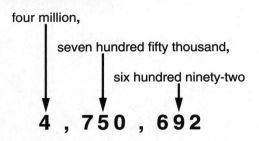

four million,
seven hundred fifty thousand,
six hundred ninety-two

4 , 7 5 0 , 6 9 2

A comma follows each period except the ones.

Practice 1

Write the standard form for the number below.

two million, six hundred eight thousand, nine hundred forty

Example 2

The numbers below are increasing each time by 1 ten thousand.

5,4**6**8,071 + 1 ten thousand
5,4**7**8,071 + 1 ten thousand
5,4**8**8,071 + 1 ten thousand
5,4**9**8,071 { 49 ten thousands +
5,5**0**8,071 { 1 ten thousand =
                { 50 ten thousands

Practice 2

Start with the number 1,906,427. Write the next five numbers, each time decreasing by 2 hundred thousands.

On Your Own

Write the word form of the number that is 3 ten thousands **more** than three million, four hundred fifty thousand, nine hundred twelve. You can start by making a place-value chart.

Write About It

Explain why these numbers have different values: 500; 5,000; 5,000,000.

Objective 3.3: Read and write whole numbers in the millions.

Lesson 3-3 — Numbers in the Millions

Try It

1. Write the word form for each number.

   a. 4,900,567 _____

   b. 732,040 _____

2. Write the standard form for each number.

   a. seven million, six hundred six thousand, forty-one _____

   b. one million, four hundred thirty-five thousand, six hundred ninety-eight _____

   c. two million, twenty thousand, twenty _____

3. Change the number 4,607,511 by each amount. Write the new number.

   a. increase by 2 ten thousands _____

   b. decrease by 3 hundreds _____

   c. decrease by 5 hundred thousands _____

4. Which is the word form of 8,005,032?

   A eight thousand, five hundred, thirty-two

   B eight million, five hundred, thirty-two

   C eight million, five thousand, thirty-two

   D eight million, fifty thousand, thirty-two

5. In 2006, the population of Dallas, Texas, was one million, two hundred eighty thousand, five hundred people. Write that in standard form.

   _____

6. What is the only seven-digit number that you could add to 8,999,999 and still have a seven-digit number? Explain.

   _____

Objective 3.3: Read and write whole numbers in the millions.

Lesson 3-4: Round Numbers through Millions

Words to Know **Rounded** numbers are values close to the original amount that are often more convenient to use.

Activity 1

Round 5,346 to the nearest thousand.
5,346 is between 5,000 and 6,000.

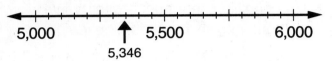

Since 5,346 is closer to 5,000 than 6,000, we round 5,346 to 5,000.

Practice 1

Round 47,206 to the nearest ten thousand.
On the number line, locate 47,206.

47,206 is closest to which ten thousand?

_____

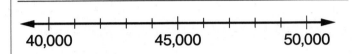

Activity 2

To round 5,234,978 to the nearest hundred thousand, underline the hundred thousands place. Circle the digit to the right.

5,2̲③4,978

Look at the circled digit.
If it is 5 or more, round up to 5,300,000.
If it is less than 5, round down to 5,200,000.

Since 3 is less than 5, round down.

Practice 2

Round 958,027 to the nearest ten thousand.

Which digit is in the ten thousands place? _____

What digit is to the right of that digit? _____

Should you round up or down? _____

What is the rounded number? _____

On Your Own

Round 542,179 to the nearest hundred thousand. Did you round up or down?

_____

_____

Write About It

Shana rounded 8,972,051 to 9,000,000. To which place(s) might she have been rounding?

Objective 3.4: Round whole numbers through the millions to the nearest ten, hundred, thousand, ten thousand, or hundred thousand.

Lesson 3-4: Round Numbers through Millions

Understand It — B

Example 1

Round 397 to the nearest ten.

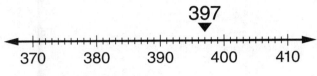

397 is closest to 400.
397 rounded to the nearest ten is 400.

Practice 1

Round 56,237 to the nearest **thousand**.
Complete the number line.

What thousand is 56,237 closest to?
Do you round up or down?

_____

Example 2

Which numbers would round to **730,000** when rounded to the nearest **ten thousand**?

73(4),602   4 is less than 5.
            Round down to **730,000**.

7(2)3,586   3 is less than 5.
            Round down to **720,000**.

72(9),428   9 greater than or equal to 5.
            Round up to **730,000**.

Practice 2

For each number, tell whether you round up or down. Then write the rounded number.

4,908,172 to the nearest **thousand**

_____

698,425 to the nearest **ten thousand**

_____

On Your Own

What is the least number you can round to the nearest million and get 2,000,000?

_____

Write About It

Ellen rounds the number 457,801 down. What is the greatest place she could have rounded to? Explain.

Objective 3.4: Round whole numbers through the millions to the nearest ten, hundred, thousand, ten thousand, or **hundred** thousand.

Lesson 3-4 — Round Numbers through Millions

1. Round each number.

   a. 4,765 to the tens _____

   b. 306,973 to the thousands _____

   c. 2,050,030 to the ten thousands _____

   d. 8,999,999 to the millions _____

   e. 68,317 to the hundreds _____

2. When you round to the nearest thousand, does the given number round to 3,540,000?

   a. 3,540,468 _____

   b. 3,540,975 _____

   c. 3,539,421 _____

   d. 3,539,539 _____

3. Show how you would use a number line to round 3,376 to the nearest thousand. Then write the rounded number.

4. Which number rounds to 6,780,000 when rounded to the nearest thousand?

   A 6,781,000   B 6,780,500

   C 6,780,972   D 6,779,500

5. Give three numbers that will round to 27,000 when rounded to the nearest thousand.

6. Sports Star baseball stadium holds 23,000 people. Tamika is in charge of ordering hamburgers and hot dogs for Saturday's game. When she places her order, would she be likely to use exact numbers or rounded numbers? Why?

Objective 3.4: Round whole numbers through the millions to the nearest ten, hundred, thousand, ten thousand, or hundred thousand.

Topic 3: Place Value beyond 1,000

Topic Summary

Circle the letter of the correct answer. Explain how you decided.

1. Last year three million, forty-seven thousand, sixteen students participated in after-school clubs. Which shows the number of students? How do you know?

   A 3,47,16

   B 3,047,016

   C 3,470,016

   D 3,470,160

   _____
   _____
   _____

2. Round 457,171 to the nearest thousand. Explain how you know.

   A 457,000

   B 458,000

   C 456,00

   D 500,000

   _____
   _____
   _____

Objective: Review skills related to place value beyond 1,000.

Topic 3: Place Value beyond 1,000

Mixed Review

1. Give the place of each underlined digit.

   a. 3<u>7</u>9 _____

   b. <u>5</u>,930 _____

   c. <u>9</u>22 _____

   Volume 1, Lesson 2-1

2. Write the standard form for six thousand, seventy.

   _____

   Volume 1, Lesson 2-2

3. Give the expanded form for each number.

   a. 3,427 _____

   b. 20,908 _____

   d. 4,902,050 _____

   Volume 1, Lesson 3-2

4. Choose the word form for 8,181

   A eighty-one thousand, eighty-one

   B eight-one, eighty-one

   C eight thousand, one hundred eighty-one

   D eighty-one thousand, eight hundred one

   Volume 1, Lesson 2-2

5. Compare with <, =, >.

   a. 194 _____ 149

   b. 723 _____ 372

   Volume 1, Lesson 1-3

6. Combine these values. Write the number in standard form.

   3 hundreds   8 millions
   0 ten thousands   1 ten
   6 hundred thousand   2 ones
   7 thousands

   _____

   Volume 1, Lesson 3-3

7. What digits could be in the blank place if the number rounded to the nearest thousand is 3,684,000?

   3,684,_29

   _____

   Volume 1, Lesson 3-4

8. Round 5,841,087 to the nearest hundred thousand.

   A 6,800,000   B 5,841,100

   C 6,000,000   D 5,800,00

   Volume 1, Lesson 3-4

Objective: Maintain concepts and skills.

Words to Know/Glossary

C

compare — Use the symbols < and > to compare two unequal quantities.

D

digits — We use ten digits to make numbers: 0, 1, 2, 3, 4, 5, 6, 7, 8, 9.

G

greatest — Highest

H

hundreds place — The hundreds place is the third place to the left of a number or a decimal point.

I

inequalities — These statements are called inequalities: 3 < 5, 5 > 3.

L

least — lowest.

M

millions period — The millions period includes the millions, ten millions, and hundred millions places.

O

ones period — The ones period includes the ones, tens, and hundreds places.

ones place — The ones place is the first place to the left of a number or a decimal point.

ordering — Writing in order or ordering means to list numbers from greatest to least or from least to greatest.

P

period — a set of three place values.

place — Each digit in a number has a place. Example, in the number 42, the digit 4 is in the tens place, and the digit 2 is in the ones place.

R

rounded — Rounded numbers have values close to the original amount that are often more convenient to use.

T

tens place — The tens place is the second place to the left of a number or a decimal point.

thousands period — The thousands period includes the thousands, ten thousands, and hundred thousands places.

V

value — The value of a digit is determined by the place it is in. Example: In 4,528, the value of the 2 is 2 tens, or 20.

W

writing in order — Writing in order or ordering means to list numbers from greatest to least or from least to greatest.

Word Bank

Word   **My Definition**   **My Notes**

Word	My Definition	My Notes

Index

C
comparing and ordering
    numbers to 100, 8–10
    numbers to 1,000, 23–25

E
eighty, 6
equivalent, 11
equivalent forms, 11–13
expanded form, 20
expanded notation, 32–34

F
fifty, 6
forty, 6

H
hundred, 6

M
Mixed Review
    1: Place Value through 100, 15
    2: Place Value through 1,000, 27
    3: Place Value beyond 1,000, 42

N
ninety, 6
numbers to 100, 5–7
    comparing and ordering, 8–10
numbers to 1,000, 17–19
    comparing and ordering, 23–25
    writing, 20–22
numbers in the millions, 35–37
    rounding, 38–40

O
ones, 2–4

P
place value to 10,000, 29–31

R
rounding through millions, 38–40

S
seventy, 6
sixty, 6
standard form, 20
symbols (<, =, >), 8–10, 23–25

T
tally marks, 12
tens, 2–4, 6
tens and ones, 2–4
thirty, 6
Topic Summary
    1: Place Value through 100, 14
    2: Place Value through 1,000, 26
    3: Place Value beyond 1,000, 41
twenty, 6

W
word form, 20
writing numbers to 1,000, 20–22

Z
zero, 32